Tell Me More About...

The Town That Brought Back the Bees

by Brittney Oden

Illustrated by
Martyna Cwiek

Ei

Earthy Info

First paperback edition August 2021

Cover Art and Illustrations by Martyna Cwiek
Cover design by Brittney Oden
Interior book design by Earthy Info

ISBN 978-1-955561-00-6 (paperback)
ISBN 978-1-955561-01-3 (ebook)
Library of Congress Control Number: 2021942039

Earthy Info
Corvallis, Oregon
www.earthyinfo.com

To my most beloved husband and son who are my
everything and encourage me to keep
working towards my dreams, support my crazy plans,
and join me on all the adventures.

To my science mentor and college professor who
took a reluctant science student and nurtured
my sense of wonder she could see in me.

Finally, to all my mentors and supporters
along the way. Who helped me to see my
inner strengths when I falter.

This book's formatting adds it to the growing list of titles
intentionally designed using The British Dyslexia Association's
style guide for easier reading. Adjustments include a carefully
selected font on cream paper, more space between the letters,
more space between the lines of text, and a jagged right edge.
Also, providing a smoother reading experience with shorter
sentences, paragraphs, and chapters. To learn more, visit the
publisher's website: www.earthyinfo.com

THE BEE LIFE CYCLE

1. THE QUEEN BEE LAYS HER EGGS AFTER MATING.

2. THE EGGS GROW... INTO LARVAE!

4. UNTIL THEY TURN INTO PUPA! ALMOST READY TO ...

3. THE LARVAE ARE FED & CARED FOR BY THE NURSE BEES.

5. BECOME WORKER BEES OR DRONES!

The Town That Brought Back The Bees

Table of Contents

Chapter 1

On the other side of town, past the school on the edge of a farm, there are hundreds of weathered wooden boxes called apiaries. The paint is chipped and the wood is rough from years of standing in the harsh hot sun. As the sun rises and warms the apiary boxes, the bees are getting ready to start their day. Many farmers living on the edge of town depend on bees to pollinate their crops.

MariAnn lives in the stacked apiaries with her colony. They have a beekeeper taking care of them. Bill the Beekeeper is a kind man. He is happiest in his garden or out on the edge of the farm in the apiary field.

One afternoon, as spring turned to summer, Bill noticed his apiaries were not as full of activity as they usually were. He looked at the field, worried he didn't see many bees pollinating the crops.

Over the next few days, the apiary boxes were almost empty. Bill searched his fields far and wide, but realized the bees were gone. They had disappeared. The queen, unhatched brood, and young bees abandoned!

There were no bee bodies anywhere, most of MariAnn's colony was simply missing.

Chapter 2

Tell Me More About. . .

Where Bees Live

Glossary

Where Bees Live

Apiaries - a place where bees are kept; a collection of beehives.

Beekeeper - a person who owns and breeds bees, especially for their honey.

Colony - animals or plants of the same kind or species. These animals or plants build a colony by living together and often times build a physical structure like a beehive.

Hive - a habitation or dwelling-place for bees, a natural habitation of bees, a crowded, busy place.

Beehives

Bees can live in many different places. Bees are busy at work in the wild. They build their beehive in trees, logs, or any empty space they can find. A beehive refers to the colony of bees that built it.

A beehive attached to a tree with sheets of honeycomb.

Up close look at bees working in their honeycomb in a hive.

Apiaries

When bees are kept and taken care of by humans they live in apiaries. Bees in apiaries still live in colonies. Their hives are kept in boxes and beekeepers care for them, use them to pollinate crops, and then harvest the honey.

Beekeepers keep wooden trays in boxes and the bees build their honeycomb inside of the trays.

A row of apiaries in an orchard.

A beekeeper working with his bees.

Chapter 3

Out past the town's edge, past any paved streets, in the middle of a mountain valley is a meadow lined with oak trees. It is the perfect place for a colony of wild bees to build their hive.

Each bee has a specific job to do. Even though there are a lot of bees, one common goal brings them together. They all work for the survival of the colony.

However, MariAnn's colony aren't the only bees busy at work. As the sun starts to rise, Nell and her sisters are busy getting ready to spend their day working among the flowers in the meadow gathering pollen.

Bill noticed that Nell's colony also started to disappear. Nell, MariAnn, and their colonies play an important role in this little agriculture town.

If the bees didn't return, the farmer's crops would stop growing. The Farmer's Market would have less fruits and vegetables, there would not be as many flowers, and there would be no honey.

Sophie's Farms!
HIVES + HERBS

OUT OF
HONEY

Cherries $10/lb

Chapter 4

Tell Me More About. . .

Pollination

Glossary

Pollination

Brood - the eggs, larvae, and pupae of honeybees.

Crops - a plant found on a farm that is grown for food, especially a grain, fruit, or vegetable.

Pollen - a fine powder, typically yellow, consisting of tiny grains.

Pollinate - to place pollen on a flower or plant, and to allow fertilization.

Queen - the single reproductive female in a hive or colony of honeybees.

Pollen

Pollen is typically yellow and is a grainy powdery substance. Each pollen grain comes from the male part of the flower or plant. There are many ways pollen travels from plant to plant. Pollen can be spread by wind, animals, or insects.

Pollen in the air released from a flower.

A jar of dried bee pollen.

Pollinate

Pollinating is the act of wind, insects, or other animals transferring pollen from one plant or flower to another.

This will cause fertilization that will later produce seeds, flowers, or fruits and vegetables.

A bee pollinating a flower while covered in pollen. Notice the yellow pollen covering the bee.

Chapter 5

The local farmers met to collaborate with others. They needed to get to the bottom of the missing bees. Their town depended on it.

A few years ago, a local scientist had visited Bill in his apiary field. He recalled that she was an entomologist and studied insects. Bill called Dr. Jasper and invited her to the meeting. The scientist explained that if MariAnn's and Nell's colonies disappeared it could change the ecosystem in their town.

They found out that people in other towns all around the world were also missing their bees. Scientists call it Colony Collapse Disorder.

Many scientists are hard at work in the labs and in the field, but they are still not sure what might be causing perfectly healthy bees to abandon their hives and colonies.

The farmers and their families became very worried. Their entire town relied on the bees to pollinate their crops. Many of the farmers tended the land their families had farmed for generations.

What would they do if the bees completely disappeared?

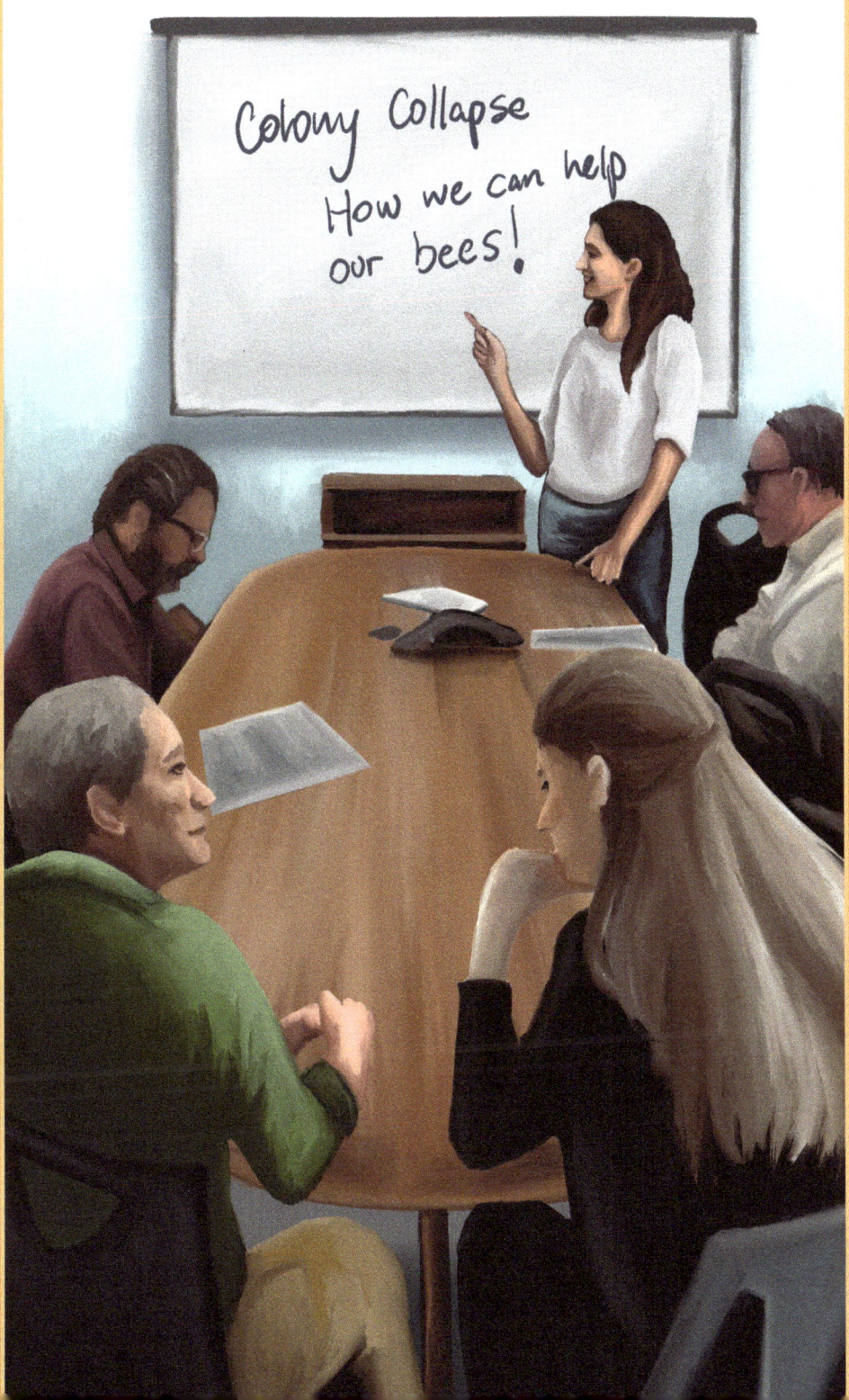

The townsfolk were anxious at the thought of their beloved agricultural town changing in ways they hadn't even imagined was possible. If the scientists didn't even know what was causing the bees to disappear, how could their town ever bring the bees back?

Seeing the doubt on their faces and hearing their hushed worried voices, Dr. Jasper quickly told them even though we may not know why this is happening, there are many things they can do to bring the bees back to their town.

Town Hall seemed to exhale as if the building had been holding its breath, just like the people inside.

Worried voices turned to excited chatter as the scientist's words filled them with hope. Dr. Jasper called for their attention and the whole room shifted forward in anticipation.

Chapter 6

Tell Me More About. . .

Colony Collapse Disorder

Glossary

Colony Collapse Disorder

Collaborate - work jointly on an activity, especially to produce or create something.

Colony Collapse Disorder - when a hive experiences a severe disappearance of worker bees in a colony that is unexpected and wipes out most of the colony.

Ecosystem - a biological community of interacting organisms and their environment.

Entomologist - a scientist who studies insects.

Colony Collapse Disorder

Colony Collapse Disorder is also called CCD. The first case of CCD was in 2006. Scientists are still unsure about what causes Colony Collapse Disorder.

There are several possible causes that include pesticides, habitat loss, infections, or viruses. But scientists can't agree on one reason why hives get CCD.

A colony suffers from CCD when they have lost most of its worker bees. The worker bees simply disappear. The queen, some nurse bees, and immature bees remain in the hive.

The hive cannot survive without the worker bees. This is not only devastating to the hive, but local crops can also die with no bees to pollinate them.

A dead bee body on the ground. With CCD bodies are not always found. Sometimes they are simply missing.

Old hive. When hives are hit with CCD they abandon their hives.

Chapter 7

Dr. Jasper explained things the town can do to support the bee population and help the hives return. Making small changes with the bees in mind would make a positive impact on the town, and help rebuild the bee's habitat.

They wrote down all of Dr. Jasper's ideas during the meeting. Everyone in town was invited to meet a month later, ready to discuss some solutions to their town's bee crisis.

Over the next few weeks, they all collected data and worked together with Dr. Jasper. They needed to figure out what had happened in their town to understand what changes they must make.

The townsfolk remembered when a lot of new people moved in several years ago. Fields full of flowers were replaced with new homes and businesses.

Nell and MariAnn's colonies had to compete for the same crops and flowers, leading to less food for both colonies.

Small and large farms both started using new pesticides, not knowing damage was being done until it was almost too late.

Certain pesticides kill all small things, both harmful and helpful bugs. Without even knowing it, the townsfolk were having a lasting impact on the environment.

They hadn't realized their sleepy little town, with a tradition and pride in agriculture, had turned into a busy town on its way to becoming a city.

Growth, while good for the town, was unhealthy for the bees.

Chapter 8

Tell Me More About. . .

Agriculture & Pesticides

Glossary

Agriculture

Agriculture - the practice of farming, which includes working the soil to grow crops and raising animals that will provide food, natural fibers, and other products.

Sustain - allow something to continue for a period of time.

Tradition - passing a culture's beliefs and customs from parents to children over many years.

Agriculture

Agriculture refers to farming and raising livestock. Farming is when farmers grow and harvest their crops. Raising livestock is when ranchers care for animals to use their meat and resources.

The agriculture industry provides us with food to eat, and other materials from natural resources. Humans need these resources to survive.

Agriculture has been around for over 100,000 years. When civilizations learned how to use agriculture they were able to grow at a rapid pace. Providing a better life for their families and the community.

However, modern agriculture also looks to science for help. Farmers and ranchers are able to use science so they can develop better seeds, improve soil quality, fertilizer, and pesticides.

With modern agriculture, humans have had both positive and negative impacts on the environment.

A tractor working in a field.

A Farmer's Market stand full of produce.

Crops on a farm.

Tell Me More About. . .

Pesticides

Glossary

Pesticides

Crops - a plant that farmers have cultivated to grow as food, especially a grain, fruit, or vegetable.

Environment - the natural world, as a whole or in a particular geographical area.

Pesticides - a liquid or granular product used to protect plants, crops and animals. Pesticides are intended to kill insects, organisms, etc that can harm plants, animals and farmer's crops.

Pesticides

Pesticides are used to control pests (harmful insects to crops) or weeds. Farmers use pesticides as protection for their crops. In larger farms pesticides are applied using a crop duster, a small plane that sprays pesticide over the crops.

Typically, people think pesticides are just chemicals. However, they can also be a virus, fungus, or bacteria.

Farmers use pesticides to protect their crops from weeds, insects, or harmful fungi that can kill off their crops.

When developing new stronger pesticides the scientists do not always understand what harmful effects these pesticides might have on the environment.

Some companies have developed pesticides that aren't as dangerous, there are also environmental friendly alternatives to pesticides.

A crop duster applying pesticides over a crop.

Unfortunately pesticides can harm helpful insects. A pile of dead bees near a crop.

Chapter 9

At their next meeting, everyone brought their ideas and put them together. The town hall was overflowing with people excited to get to work and support the bees. Teachers even brought their classes with projects they researched.

As the meeting came to a conclusion, they decided to run a town campaign encouraging everyone to plant flowers wherever they could.

The town would plant bee gardens in town parks and community areas. The farmers agreed to find bee friendly pesticides. The students planned to create flyers and bring awareness to the town's "Bring Back the Bees" initiative.

Everyone left the meeting that night with a sense of hope and community.

Later that summer Bill stood in the hot sun looking over his fields. A big smile started to spread across his face. The kind of smile that reaches your eyes. MariAnn, Nell, and their colonies had returned.

Chapter 10

Tell Me More About. . .

Community Science

Glossary

Community Science

Awareness - knowledge that something exists, such as a problem. Understanding of a situation based on information or experience.

Campaign - an organized course of action to achieve a goal.

Initiative - a group of people or organization creating a strategic plan with the intent to resolve a problem or challenge. Thinking outside the box to develop a solution.

Community Science

Community science is sometimes called Citizen Science. But they both mean the same thing.

Community science is when regular people volunteer to help scientists with experiments, collect or analyze data, etc.

Originally when various science fields were being explored and discoveries were being made most of the work was done by explorers or wealthy gentlemen who devoted their lives to science. Later, only findings from scientists with degrees were taken seriously.

That is until the movement of community science changed everything. Scientists realized that using people in the community was a valuable asset. Scientists and community members now work together to further science.

There are many apps, projects, and foundations to get involved with to improve your community. You could participate in or host your own BioBlitz to catalog data in a specific area. Check it out today!

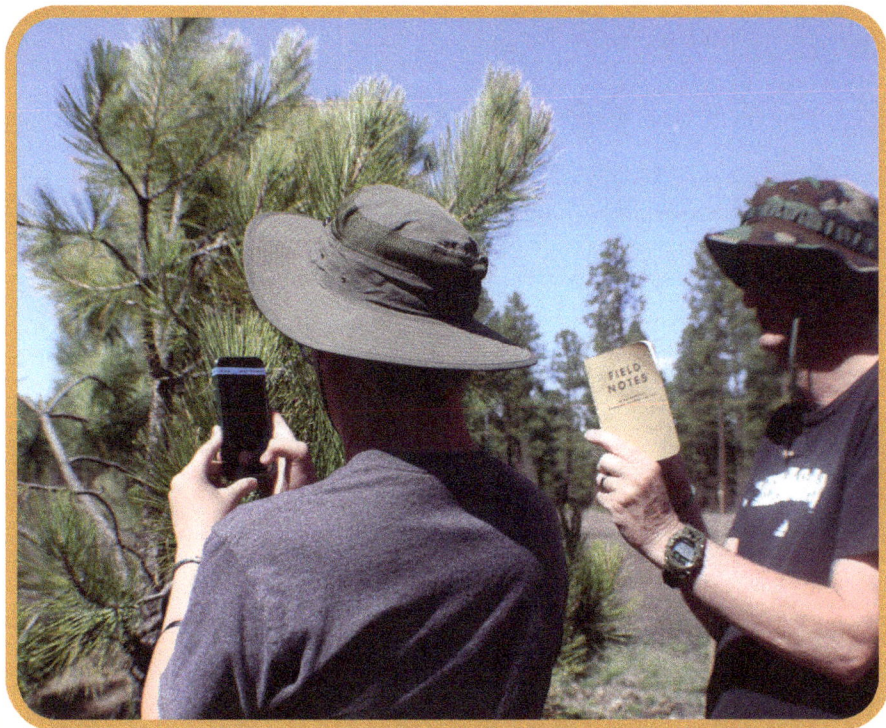

Download an app and help collect data in nature to help scientists track species or the change of habitats over time.

During a BioBlitz community members use a variety of tools to find and identify different species.

Teacher Guide

Educators, this is a great resource for your classroom to encourage and engage in an inquiry-based teaching model. Mentor texts are a great way to launch problem-based learning in your classroom. Read this book as an introduction to pollinators and the dangers they face.

Problem-Based Learning Outline

Intro – Read "The Town That Brought Back the Bees" asking guiding questions noted below.

Reading Guide

Front Cover
Read the title, show the cover and ask the class what they think the book is about.

Chapter 1 & 3
After reading, ask the class: "What do you think their important role is?", "Where do you think the bees have gone?", and "What will happen with no bees?"

Chapter 5
"What small changes do you think the town could make?"

Chapter 7

After reading ask "Have you seen changes in your community?" and "What effects do you think these changes have had?"

Chapter 9

Ask the class what ideas they have to bring the bees back.

End of book

Let the class know that you will be exploring what is happening to bees in your area, region, etc. and will be brainstorming what you can do to bring back the bees to your area. After the research is complete you will decide as a class which solution(s) to the problem they can implement.

Problem

Depending on grade level and student abilities have students brainstorm what the problem is. One essential question you can use is "How Can We as a Class Support the Bee Population In our Community?" PBLWorks has a great tool called a "Tubric" to help develop meaningful, and real-life world problems for problem-based learning questions.

Research

After developing your class essential question have students research problems facing bees. Such as Colony Collapse Disorder, loss of habitat, and pesticide use. Try to bring in local experts, find any bee sanctuaries or bee/butterfly gardens in your area, etc. Do observations on campus, nearby parks or nature areas, etc.

Solution/Presentation

Have students break up in groups, then have students research solutions to these problems. The key to PBL is that there is no one answer to the problem or one way to present their findings. Students and educators are limited only by your imagination. Arrange for the class to present to other classes, parents, do this project with a school in another district or another school within your district and present to each other over an online platform such as Zoom, Webex, etc.

Action

After the presentations decide what solution(s) to implement and help students to carry out their solutions.You could even get your school involved at all grade levels and run your own "Bring Back the Bees" initiative.

Brittney Oden is an educator who is passionate about science and making it accessible to all diverse learners. While teaching junior high special education she saw a need for science content at a reading level her students could engage with. She also trains educators and presents at conferences to change the mindset that science is an "extra" and shows how to implement reading, writing, and math within science. In her spare time she loves to go on adventures with her family and has fun being a nature photographer.

Martyna Cwiek is a freelance illustrator and designer living in an Ontario suburb with her husband, dog and guinea pigs. She spends her free time drawing, reading science fiction and poorly attempting to grow vegetables.

References and Resources for Further Reading

Hladik, Michelle. Native Bees are Exposed to Neonicotinoids and Other Pesticides. USGS.

Meisset, Beatriz, and Stephen Buchmann. Bee Basics: An Introduction to Our Native Bees. Washington, DC: USDA Forest Services, 2011.

Rosner, Hillary. Return of the Natives: How Wild Bees Will Save Our Agricultural System. Scientific American.

Wilson-Rich, Noah. The Bee: A Natural History. Princeton, NJ: Princeton University Press, 2014.

Jump into Community Science for bees with Bumble Bee Watch at BumbleBeeWatch.org

Learn more about bees and honey with the National Honey Board at honey.com

See what the EPA is doing to help at epa.gov/pollinator-protection

www.ingramcontent.com/pod-product-compliance
Lightning Source LLC
Chambersburg PA
CBHW052026030426
42335CB00026B/3301